# Tony Hunt's Structures Notebook

# Tony Hunt's Structures Notebook

Tony Hunt

## Architectural Press

AMSTERDAM  BOSTON  HEIDELBERG  LONDON  NEW YORK  OXFORD  PARIS

SAN DIEGO  SAN FRANCISCO  SINGAPORE  SYDNEY  TOKYO

Architectural Press
An imprint of Elsevier
Linacre House, Jordan Hill, Oxford OX2 8DP
30 Corporate Drive, Burlington, MA 01803

First edition 1997
Reprinted 1998
Second edition 2003
Reprinted 2005

**British Library Cataloguing in Publication Data**
A catalogue record for this book is available from the British Library

ISBN 0 7506 5897 5

For information on all Architectural Press publications
visit our website at www.architecturalpress.com

Working together to grow
libraries in developing countries

www.elsevier.com | www.bookaid.org | www.sabre.org

ELSEVIER    BOOK AID International    Sabre Foundation

Transferred to digital print 2007

Printed and bound by CPI Antony Rowe, Eastbourne

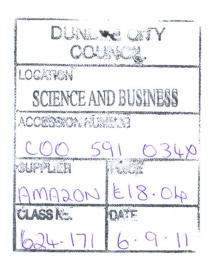

# Contents

Preface to the First Edition     vii

Preface to the Second Edition     ix

1    Introduction     1

2    Structure and structural form     2

3    Structural materials     12

4    Loads on structure     20

5    Equilibrium     30

6    Structural elements and element behaviour     40

7    Structural types     50

8    Some further significant structures and assemblies     80

Appendix I    Tensile strength of some common materials     95

Appendix 11    Bending and deflection formulae for beams     98

Appendix III    Reading list     99

# Preface to the First Edition

The *Structures Notebook* was originally written by Tony Hunt as a brief teaching aid for students at the Royal College of Art who had very little, if any, knowledge of physics or structural behaviour. The original *Notebook* was oversimplified but served its purpose as a primer. It has now been expanded into a more comprehensive book while retaining a simple visual and non-mathematical approach to structural behaviour.

The purpose of the *Structures Notebook* is to explain, in the simplest possible terms, about the structure of 'things', and to demonstrate the fact that everything you see and touch, live in and use, living and man-made, has a structure which is acted upon by natural forces and which reacts to these forces according to its form and material.

The book is divided into seven main sections, in a logical sequence, and is written in simple language. Each section, related to its text, contains a comprehensive set of hand-drawn sketches which show, as simply as possible, what the text is about. The book is almost totally non-mathematical, since the author believes very strongly that structural behaviour can be understood best by diagrams and simple descriptions and that mathematics for the majority of people interested in design is a barrier. The design of structures is a combination of art and science and to achieve the best solution, concept should always come before calculation.

*Professor Tony Hunt*

# Preface to the Second Edition

Since this book is about the basics of structure and structural behaviour, both of which are subject to the laws of physics and mathematics, it is difficult to know what to add.

This book was first published in 1997. The author feels that some key structural ideas and assemblies were not illustrated and they have now been included, together with some recent examples. These expand the range of ideas conceived by different designers and show further different ways of creating inventive structures and structural assemblies. These illustrations are included in a new Chapter 8.

Recently, with some designers, there has been a move away from orthogonal geometries to more random forms (see 'informal' by Cecil Balmond). This has been aided by the enormous power of modern analytical and graphic computing, and has been driven by both architects' and engineers' interest in exploring more complex geometries. This adds to the complexity of design solutions and has to be considered as part of current design thinking.

Finally, the reading list has been added to for up-to-date references to books which I consider to be important for architects, engineers and designers.

*Professor Tony Hunt*

# 1

# Introduction

This book is about the basic structure of things. Its aim is to develop an understanding of essential structural principles and behaviour by a descriptive and largely non-mathematical approach. It relates to the structure occurring in such diverse objects as a bridge, a box for packaging, furniture, buildings etc. and it covers all the common structural elements singly and in composite form.

This book is a primer on the subject. There are a large number of books on building structures, the most important or relevant of which are listed in Appendix III.

# 2

# Structure and structural form

## Structure

### What structure is

Structure is the load-carrying part of all natural and man-made forms. It is the part which enables them to stand under their own weight and under the worst conditions of externally applied force.

### The designer

In the context of structure, a designer is one who conceives a structural part or a structural system which functions satisfactorily, is integrated successfully within the overall design and is appropriate for its purpose in terms of material and form.

## The design process

Without a brief it is not possible to design, since there are no rules and no constraints. Therefore, no matter how sketchy, it is the brief which sets the basic framework for the designer. It provides the lead-in for the first analysis of the problem which then develops into an iterative process, with ideas being tested, modified, rejected, until an appropriate solution to the problem is reached.

## Optimum design

A designer should generally aim for the optimum solution in order to obtain the maximum benefit with the minimum use of material within the constraints of strength, stiffness and stability. The result will be EFFICIENCY combined ideally with ELEGANCE AND ECONOMY.

## Influences on the designer

The major influences on creative structural design are:

| | | |
|---|---|---|
| *Precedent* | – | what's gone on |
| *Awareness* | – | what's going on |
| *Practicality* | – | how to do it |

# Structural form

Structures take one of four basic forms which may exist singly or in combination.

*Solid*    An homogeneous mass structure where the external surface is independent of the internal form – a three-dimensional solid body

*Surface*   An homogeneous surface where the external and internal forms are similar – a two-dimensional panel

*Skeletal*   A framework where the assembly of members gives a clear indication of the form usually using one-dimensional elements

*Membrane* A flexible sheet material sometimes reinforced with linear tension elements used either as single cables or as a cable net. A variation is the pneumatic where air under pressure is contained by a tension membrane skin

*Hybrid*   A combination of two of the above forms of near equal dominance

For examples of all the above, see Chapter 7.

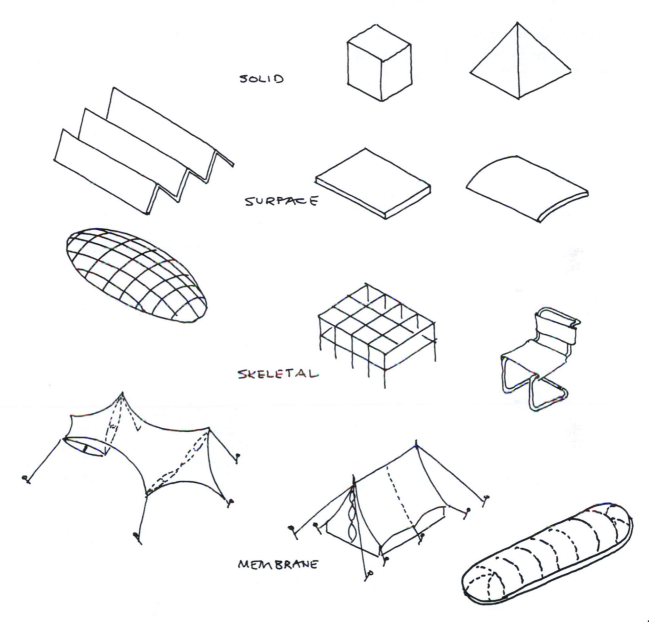

SOLID

SURFACE

SKELETAL

MEMBRANE

# Structural form in nature

Here are some examples of objects in nature,
all of which have a structure in one or more
forms:

*Human and animal skeletons*
*Birds' wings*
*Fish*
*Flowers*
*Honeycombs*
*Leaves*
*Plants*
*Rock caves*
*Shellfish*
*Snails*
*Snowflakes*
*Spiders' webs*
*Trees*

CAVE

SPIDER WEB

FISH SPINE

HONEY COMB

NAUTILUS

DAFFODIL

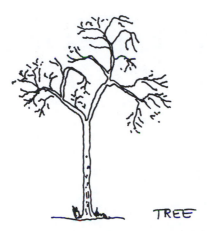

TREE

GRASS HOPPER

FIG LEAF

BIRD WING

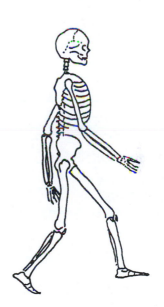

SKELETON

## Structural form – man-made

Here are some examples of man-made objects, all of which have a structure in one or more forms:

*Aeroplanes*
*Bicycles*
*Bridges*
*Buildings*
*Cars*
*Clothes*
*Cranes*
*Dams*
*Engines*
*Fabrics*
*Fastenings*
*Furniture*
*Musical instruments*
*Packaging*
*Pottery*
*Roads*
*Sculpture (3-D art)*
*Ships and yachts*
*Sports gear*
*Technical instruments*
*Tents*
*Tools*
*Toys*
*Tunnels*
*Wheels*

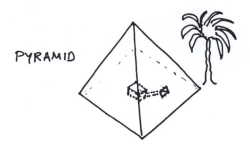

PYRAMID

BICYCLE

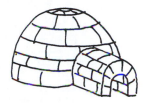

IGLOO

WIGWAM

TWIG + BARK
HUT

MICRO ELECTRONICS FACTORY

RIETVELD CHAIR

9

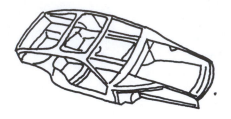

CAR BODY - MONOCOQUE

PALM HOUSE KEW

MICROLITE
SINGLE SEAT
AIRCRAFT

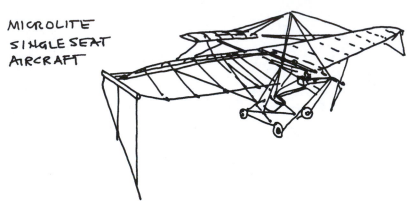

FORTH RAIL BRIDGE

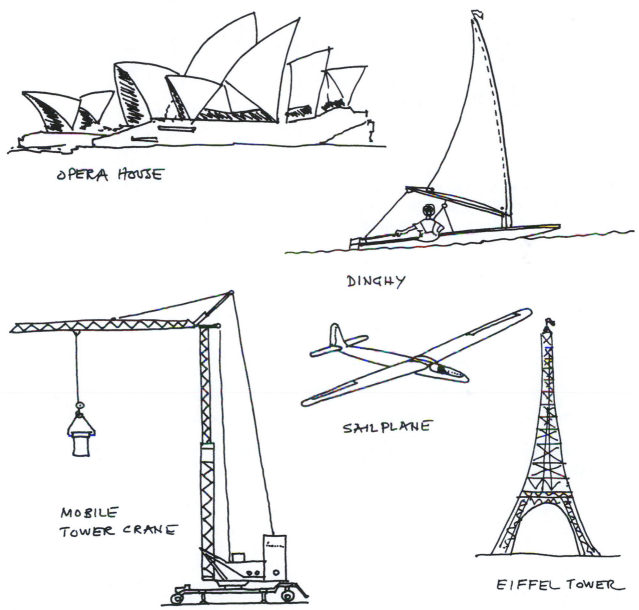

OPERA HOUSE

DINGHY

MOBILE
TOWER CRANE

SAILPLANE

EIFFEL TOWER

# 3

# Structural materials

All materials have a stiffness and strength and are manufactured into a shape. Stiffness and strength are different, complementary characteristics and describe the properties of a solid material. Shape affects performance.

## Strength

Strength is the measure of the force required to break the material.

A material can be strong or weak – see Appendix I.

| | |
|---|---|
| *Mild steel* | stiff and strong |
| *Sheet glass* | stiff and weak |
| *Nylon rope* | flexible and strong |
| *Rubber* | flexible and weak |

# Stiffness

The majority of structural materials behave in an elastic manner according to Hooke's Law which states that elastic extension is proportional to load. When the load is removed, the material recovers its original length and shape.

Different materials have different stiffness characteristics. They can be: stiff, flexible, stretchy, springy or floppy.

This stiffness is defined for each material as the *E*-value – Young's modulus, named after its discoverer.

*E* is the value of stress/strain and is a constant for a given material.

Stiffness and strength do not necessarily go hand in hand as the above examples show.

# Shape

Shape is the third property which affects the performance of a material in a particular loading situation. In pure tension, shape does not matter, but in all other loading modes – compression, bending and shear – the cross-sectional shape affects performance.

In general terms, for maximum performance, the material should be arranged in order to be as far away from the centre of the section as possible.

# Material behaviour

Materials are either 'isotropic' or 'anisotropic' depending on their behaviour under load.

*Isotropic*    Providing equal performance in all directions in both tension and compression

*Anisotropic*    Providing differing performances in different directions and in compression and tension

Some examples:

## Isotropic materials

### Metals
Including steel, aluminium, bronze, titanium etc.

**Anisotropic materials**

## Timber
Different values for compression and tension. Different values for load parallel and perpendicular to the grain.

## Concrete and masonry
Good in compression, poor in tension. Steel reinforcement provides the tension element in reinforced concrete.

## Plastics and reinforced plastics
Usually stronger in tension than compression. A very wide range of performance according to type of plastic and reinforcement.

Concrete - insitu + precast

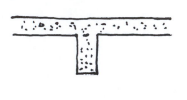

## STRUTS

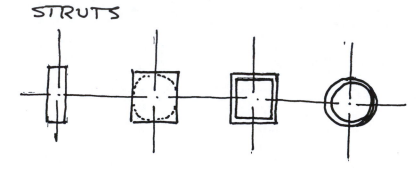

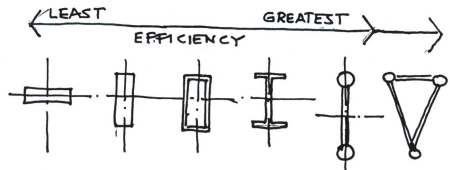

## BEAMS

## COMMON STRUCTURAL MATERIAL SHAPES

Hot rolled steel
Extruded Aluminium

Note - Aluminium can be extruded through most die shapes

Cold formed steel

Timber

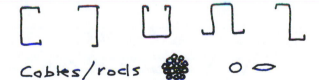

Cables/rods

Masonry - cut to squared stone or pressed + baked as bricks

Concrete - cast to shape in formwork moulds

17

## COMPOSITE MATERIAL PANELS

Aluminium skins +
rigid plastics foam core

Aluminium skins + phenolic resin
coated paper core (Nomex)

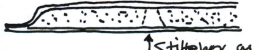

↑ stiffener as required

Injection moulded GRP
with rigid isocyanate core

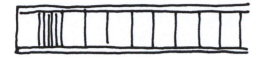

GRP skins on end-grain
Balsa core

Vacuum-formed superplastic
aluminium with rigid foam core

a air supply

ETPE foil in aluminium frame
Air-inflated

# 4

# Loads on structure

All structures develop internal forces which are the result of external applied loads and the weight of the structure itself.

Loads are conventionally divided into a number of classifications under the following headings:

**Permanent**

*Dead load*  The self-load of the object or part due to its mass

**Temporary**

*Imposed load*  The 'user' load which is removable and thus is a 'live' load

*Thermal load*  The load induced by temperature change causing expansion or contraction of the object

*Dynamic load*  A cyclical load caused by varying external conditions which cause the object to vibrate or oscillate

**Structures must always be designed for the worst anticipated combination of loading otherwise unserviceability or failure can result.**

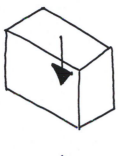

DEAD LOAD

IMPOSED LOAD

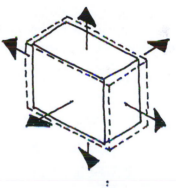

THERMAL LOAD

DYNAMIC LOAD
ON/OFF

# Examples of load cases

## Dead load

*Aeroplane*     The weight of the plane without fuel, passengers or baggage

*Building*     The weight of the structure, cladding, fixed equipment etc.

*Vehicle* (bus, railway carriage, truck, passenger car)     The weight of the vehicle without fuel, passengers or freight

*Yacht*     The unladen weight of the vessel

*Object* (e.g. chair)     The weight of the chair itself

All these are examples of permanent load.

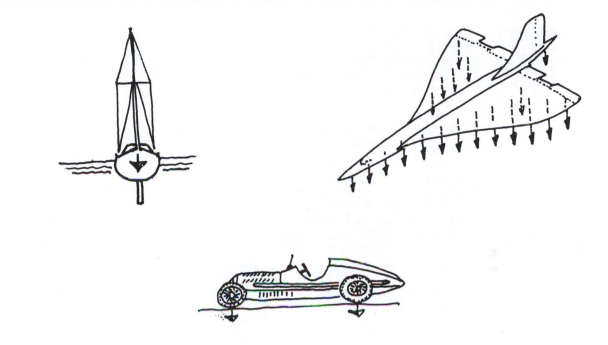

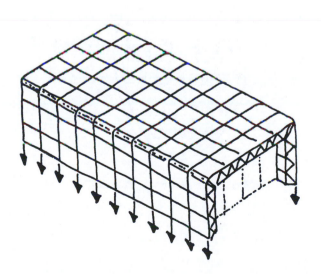

# Imposed load

*Aeroplane*  The fuel, passengers and cargo all of which are variable

*Building*  The 'user' load – people, furniture, factory machinery, any equipment which is movable. Environmental loads – snow, the 'static' effects of wind

*Vehicle*  The fuel, passengers, freight etc.

*Yacht*  The crew, stores, fuel, water etc.

*Object* (chair)  A person sitting or standing on or tilting a chair

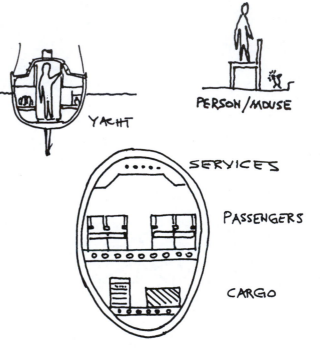

YACHT

PERSON/MOUSE

SERVICES

PASSENGERS

CARGO

AEROPLANE FUSELAGE

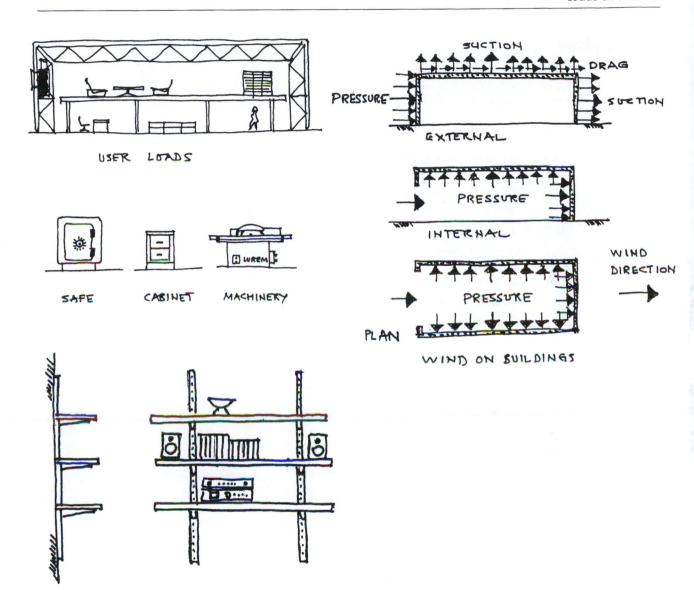

USER LOADS

SAFE     CABINET     MACHINERY

SHELF UNIT

SUCTION
DRAG
PRESSURE
SUCTION
EXTERNAL

PRESSURE
INTERNAL

PRESSURE
PLAN

WIND DIRECTION

WIND ON BUILDINGS

25

# Thermal load

*Aeroplane*   Temperature changes in the skin due to height and speed (speed causes air friction and generates heat)

*Building*   Roofs and walls facing the sun are subject to diurnal temperature change. Elements may have a different outside and inside temperature

*Vehicle*   The engine increases in temperature due to combustion and outside air temperature. It requires cooling

*Object*   A hot liquid poured into a glass can cause it to shatter – thermal shock. A spoon in the glass acts as a 'heat sink'

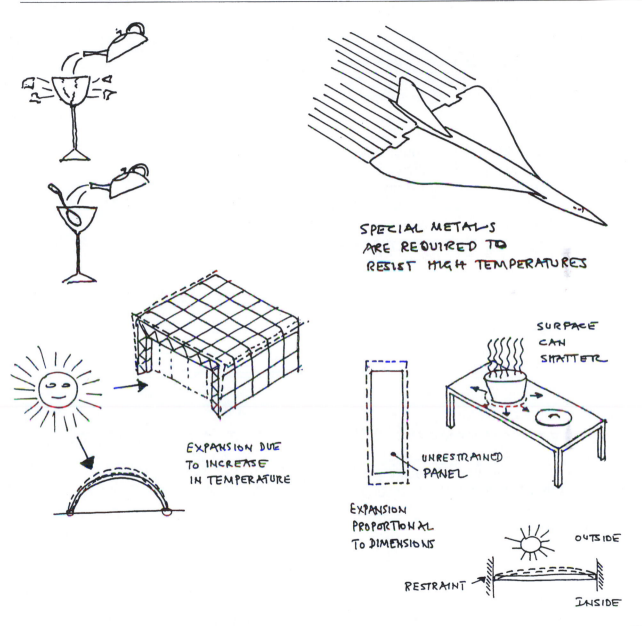

SPECIAL METALS ARE REQUIRED TO RESIST HIGH TEMPERATURES

EXPANSION DUE TO INCREASE IN TEMPERATURE

SURFACE CAN SHATTER

UNRESTRAINED PANEL

EXPANSION PROPORTIONAL TO DIMENSIONS

RESTRAINT

OUTSIDE

INSIDE

27

# Dynamic load

*Aeroplane*   A sudden change in direction causes dynamic flexing of the wings and G forces on humans

*Building*   Gusty wind conditions cause oscillations. Surge caused by a lift starting and stopping. Surge due to overhead crane travel

*Vehicle*   Accelerating, decelerating and cornering all cause dynamic loads on parts of the vehicle

*Yacht*   Wind on the sail causing heel (overturning). 'Pounding' of the hull in heavy seas

*Bridge*   Rolling loads cause the bridge deck to flex

*Object*   Rocking or tilting a chair is dynamic and affects the joints

Imposed, thermal and dynamic loads are all temporary loads but their worst combination added to the dead load must be considered for design purposes.

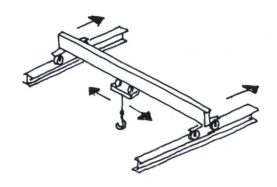

OVERHEAD TRAVELLING CRANE
– CREATES SURGE

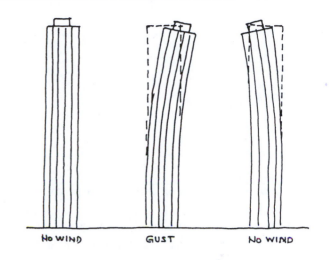

WIND INDUCED OSCILLATION ON A TOWER

No Wind     GUST     No WIND

DROP INTO AIR POCKET CAUSES EXTRA LOAD
– WINGS FLEX

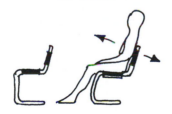

CHAIR FLEXES

WIND

YACHT HEELS

DYNAMIC LOADING CAUSES
FLEXING OF CABLES AND BRIDGE DECK

⊗    STOCKBRIDGE DAMPER

# 5
# Equilibrium

To stand up and stay in place structures must be in equilibrium.

External loads act on a structure and induce internal forces, both loads and forces having magnitude and direction.

For equilibrium, reactions must act in an equal and opposite sense to the applied loads.

There are three conditions which may have to be satisfied to achieve equilibrium depending on the form of loading. These conditions are expressed as simple equations with meanings as follows:

$V = 0$     The sum of vertical loads and reactions must equal zero

$H = 0$     The sum of horizontal loads and reactions must equal zero

$M = 0$     Clockwise moments must equal anti-clockwise moments

Moment = load × distance of load from support or point of rotation

P  =  Load (linear)
R  =  Reaction (linear)
M  =  Moment (bending or rotation)
Σ  =  Sum

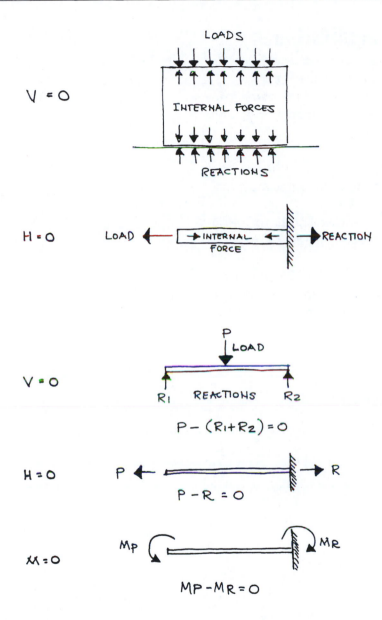

V = O

LOADS

INTERNAL FORCES

REACTIONS

H = O

LOAD — INTERNAL FORCE — REACTION

V = O

P
LOAD

REACTIONS

$R_1$        $R_2$

$$P - (R_1 + R_2) = 0$$

H = O

P ← → R

$$P - R = 0$$

M = O

$M_P$        $M_R$

$$M_P - M_R = 0$$

# Examples of equilibrium

*Vertical*    The load and reactions of an object sitting on the floor
A horizontal structure carrying a vertical load producing end reactions

*Horizontal*    The tug of war where, for equilibrium, both teams must pull with equal force.
The vehicle travelling horizontally which meets an obstruction

*Rotational*    The see-saw where the sums of the loads x their distance from the point of support must equal for balance

Notes:
$\Sigma$ = sum
$\neq$ = not equal

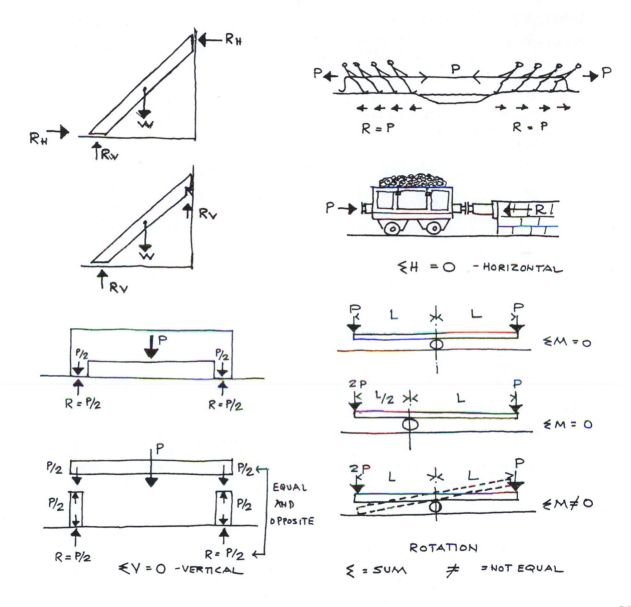

$R_H$

$R_H$ → $\uparrow R_V$

$W$

$R_V$

$W$

$\uparrow R_V$

P ← $\quad$ → P ← P $\quad$ → P

R = P $\qquad$ R = P

P → $\quad$ ← R

$\xi H = 0$ — HORIZONTAL

$\downarrow P$

P/2 $\quad$ P/2

R = P/2 $\qquad$ R = P/2

P

P/2 $\quad$ P/2

P/2 $\quad$ P/2

P/2 $\quad$ P/2

R = P/2 $\qquad$ R = P/2

EQUAL
AND
OPPOSITE

$\xi V = 0$ — VERTICAL

P $\quad$ L $\quad$ L $\quad$ P

$\xi M = 0$

2P $\quad$ L/2 $\quad$ L $\quad$ P

$\xi M = 0$

2P $\quad$ L $\quad$ L $\quad$ P

$\xi M \neq 0$

ROTATION

$\xi$ = SUM $\qquad \neq$ = NOT EQUAL

33

# Equilibrium

*W* is constant for object

As *P* increases, rotation increases

When action line of *W* falls outside base then object is unstable and falls over

*For stability*

$$P_y < W_x$$

If *y* increases and *x* decreases, instability occurs when $P_y > W_x$

*Graphical solution*
*P* and *W* are drawn to scale to represent load magnitude and direction, *R* is resultant

| | | |
|---|---|---|
| *P* | = | external load |
| *W* | = | weight of object |
| *R* | = | reaction |
| > | | greater than |
| < | | less than |

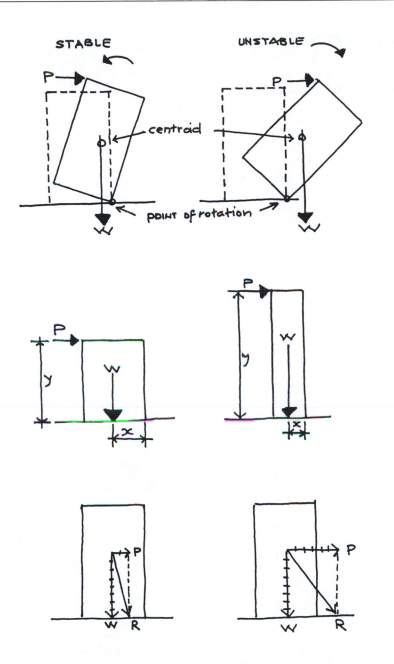

STABLE

UNSTABLE

P

centroid

P

point of rotation

W

W

P

y

W

x

P

y

W

x

W   R

P

W   R

P

# Parallelogram of forces (vectors)

Forces acting at a point in a direction other than vertical or horizontal can be resolved into vertical and horizontal by a *vector diagram drawn to scale* or by trigonometry. Similarly, two forces can be resolved into a resultant by the same method. The opposite of the resultant forms the equilibrium force.

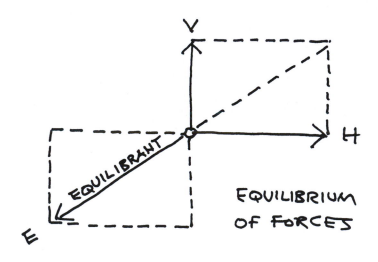

EQUILIBRIUM OF FORCES

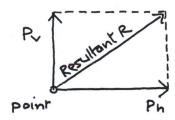

$P_v$  Resultant R  point  $P_h$

2 Forces at Rt. angles
1 Resultant

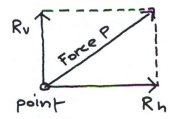

$R_v$  Force P  point  $R_h$

1 Force
2 Resultants at Rt. angles

TRIANGLE OF FORCES

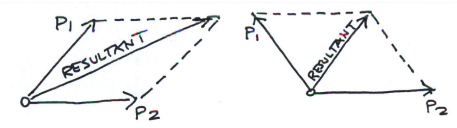

$P_1$  RESULTANT  $P_2$        $P_1$  RESULTANT  $P_2$

PARALLELOGRAM OF FORCES

# Settlement and earthquake behaviour

Both settlement and earthquakes can cause movements and distress in building structures.

Settlement occurs due to compression of the soil under the foundations. Differential settlement occurs due to uneven bearing capacity of the soil or to uneven loading.

Earthquakes give rise to horizontal ground movement and can also be the cause of settlement due to ground compaction.

TOWER OF PISA
- COMPRESSIBLE SOIL LAYER ON INCLINE
- UNEVEN SETTLEMENT

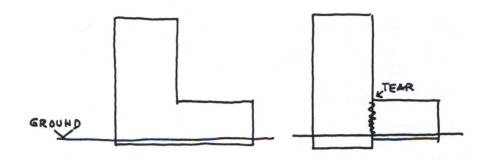

UNEVEN LOADING - DIFFERENTIAL SETTLEMENT
AND DAMAGE

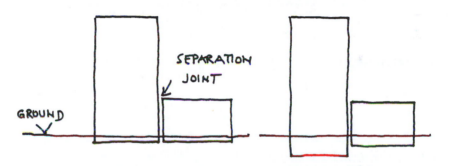

UNEVEN LOADING - DIFFERENTIAL SETTLEMENT
NO DAMAGE

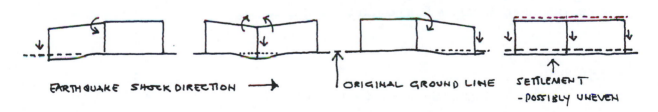

EARTHQUAKE SHOCK DIRECTION ⟶          ORIGINAL GROUND LINE          SETTLEMENT
- POSSIBLY UNEVEN

EARTHQUAKE BEHAVIOUR - SINGLE STOREY FRAME BUILDING

# 6

# Structural elements and element behaviour

## Structural elements

The design of a structural element is based on the loads to be carried, the material used and the form or shape chosen for the element. See Chapter 7.

The elements from which a structure is made or assembled have, in engineering or building terms, specific names which are used for convenience. In other disciplines such as naval architecture and furniture design the names are different but the functions are the same.

## The elements

*Strut*      A slender element designed to carry load parallel to its long axis. The load produces compression

| | |
|---|---|
| *Tie* | A slender element designed to carry load parallel to its long axis. The load produces tension |
| *Beam* | Generally a horizontal element designed to carry vertical load using its bending resistance |
| *Slab/plate* | A wide horizontal element designed to carry vertical load in bending usually supported by beams |
| *Panel* | A deep vertical element designed to carry vertical or horizontal load |

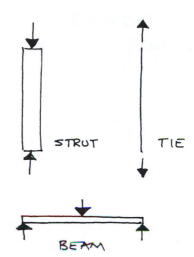

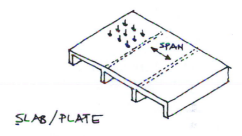

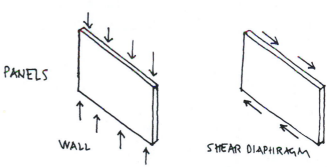

**41**

# Element behaviour – deformation

The loaded behaviour of structural elements is dependent on internal and external factors.

*Internal factors* – type of material, cross-sectional shape, length, type of end fixity

*External factors* – type of position and magnitude of load

Under load, elements deform in the following ways:

Struts compress under load and can buckle if not stabilized laterally

Ties extend under load

Beams and slabs deflect due to bending

Panels deform due to in-plane load

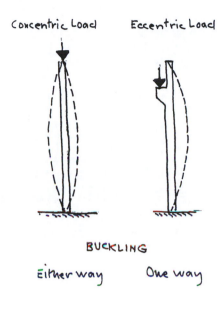

Concentric Load    Eccentric Load

BUCKLING

Either way    One way

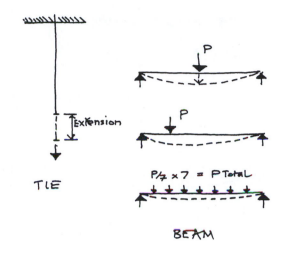

TIE

Extension

P

P

P/7 x 7 = P Total

BEAM

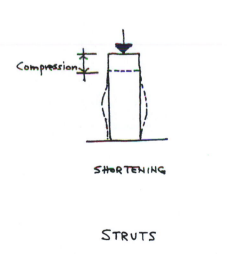

Compression

SHORTENING

STRUTS

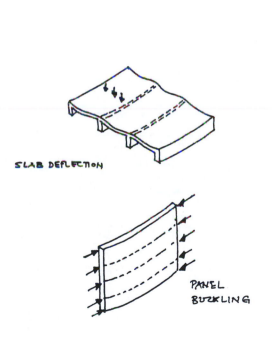

SLAB DEFLECTION

PANEL BUCKLING

43

## DEFORMATION UNDER LOAD

### BEAMS

pin ends

built in ends
– encastré

cantilever

cantilever
/pin end

Propped
cantilever

continuous 2-span beam

### COLUMNS

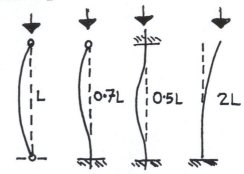

L    0.7L    0.5L    2L

Effective Length / End Fixity

FRAMES

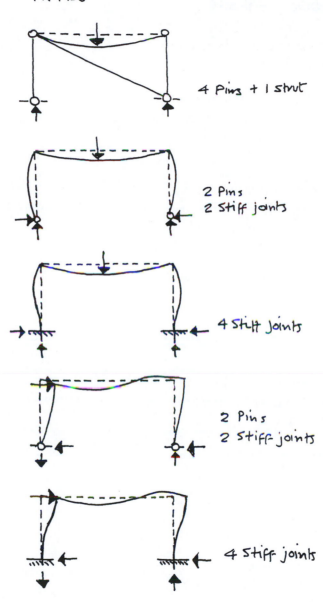

4 Pins + 1 strut

2 Pins
2 stiff joints

4 Stiff Joints

2 Pins
2 stiff joints

4 stiff joints

45

# Element behaviour – stress

When an element is loaded it becomes stressed. The type of stress and its effects on an element are as follows:

*Tensile*      The particles of material are pulled apart and the element increases in length. A tie is in tension

*Compressive*    The particles of material are pushed together with a consequent decrease in length. A strut is in compression

*Shear*      The particles of material slide relative to one another

*Torsion*     A form of shear caused by twisting

*Bending*    A combination of tension, compression and shear. Beams are in bending

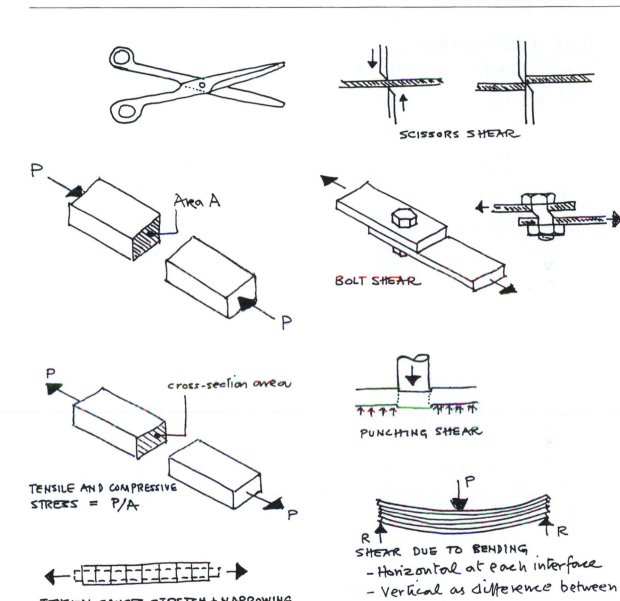

SCISSORS SHEAR

P

Area A

P

BOLT SHEAR

P

cross-section area

P

TENSILE AND COMPRESSIVE
STRESS = P/A

PUNCHING SHEAR

P

R                                    R

SHEAR DUE TO BENDING
- Horizontal at each interface
- Vertical as difference between
  P and R

TENSION CAUSES STRETCH + NARROWING

# Some stress and strain definitions

Stress $\quad\quad\quad$ $f\;$ = load per unit area = $P/A$

Ultimate stress $\quad$ $fu$ = the stress at which a
$\quad\quad\quad\quad\quad\quad\quad\quad\quad$ material fails

Working stress $\quad$ $fw$ = the safe maximum stress
$\quad\quad\quad\quad\quad\quad\quad\quad\quad$ for a material

Strain $\quad\quad\quad\quad$ $e\;$ = extension per unit length
$\quad\quad\quad\quad\quad\quad\quad\quad\quad$ under load

$\dfrac{\text{Extension}}{\text{Original length}}$ $e\;$ = $1/L$

Modulus of $\quad\quad$ $E\;$ = a constant defining the
elasticity $\quad\quad\quad\quad\quad\quad$ stiffness of a material

$\quad\quad\quad\quad\quad\quad$ $E\;$ = $\dfrac{\text{stress}}{\text{strain}}$

Factor of safety $\quad$ = $\dfrac{\text{Ultimate stress}}{\text{Working stress}}$

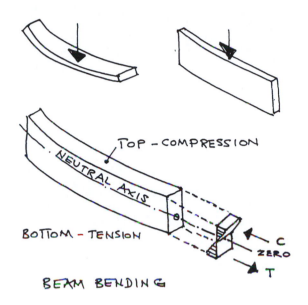

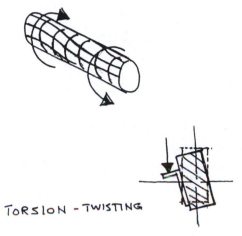

TOP – COMPRESSION

NEUTRAL AXIS

BOTTOM – TENSION

C

ZERO

T

**BEAM BENDING**

**TORSION – TWISTING**

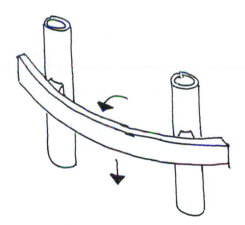

**COMBINED BENDING + TORSION**

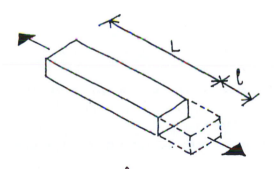

L

ℓ

$$STRAIN \ e = \frac{\ell}{L}$$

TENSILE + COMPRESSIVE STRAINS
ARE SIMILAR

SHEAR + TORSIONAL STRAINS
ARE MORE COMPLEX

49

# 7

# Structural types

Structures can be classified by their basic forms.

*Solid*      Walls, arches, vaults, dams etc.

*Surface*    Grids, plates, shells, stressed skins

*Skeletal*   Trusses and frameworks

*Membrane*  Cable/membrane tents, cable nets, pneumatics

*Hybrids*    Tension-assisted structures

The classifications are not mutually exclusive. For example a thin curved shell dam would be classified as a surface structure.

Combinations of more than one type are common. For example, skeletal frameworks are often stiffened by the insertion of a panel which is a surface structure. Buildings and furniture, aircraft and vehicles are treated in this way. Similarly, monocoque structures are a combination of skin and skeletal.

# Walls, arches and vaults

Walls are the simplest form of compression structure with loads transmitted vertically downwards. Construction is usually in masonry or concrete. When stiffened with ribs they can also act as retaining structures.

Arches and vaults carry compression loads in a most efficient way due to their curvature. Construction traditionally is in masonry, more recently in reinforced concrete.

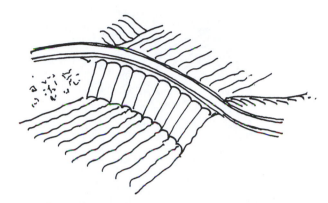

CONCRETE VAULT DAM

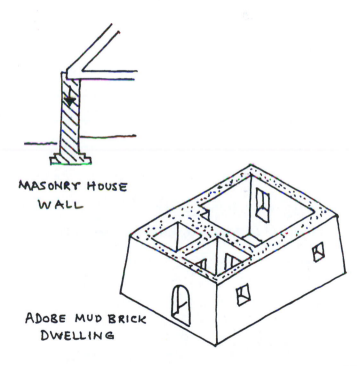

MASONRY HOUSE WALL

ADOBE MUD BRICK DWELLING

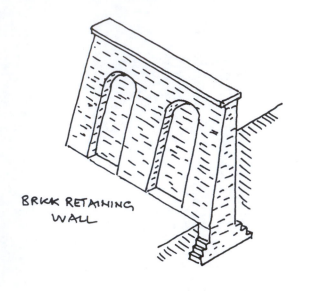

BRICK RETAINING
WALL

BRICK AQUEDUCT ARCHES

STONE BRIDGE

ROMANESQUE ARCH

FINIAL

STONE VAULT

ARCH BRIDGE
(strictly a frame structure)

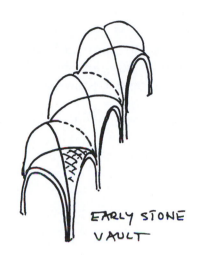

EARLY STONE
VAULT

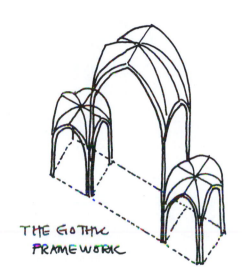

THE GOTHIC
FRAMEWORK

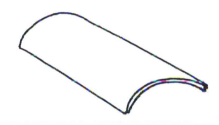

BARREL VAULT

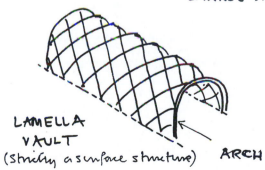

LAMELLA
VAULT
(strictly a surface structure)

ARCH

# Trusses

Trusses are an assembly of structural members based on a triangular arrangement with member to member pin-jointed connections called 'nodes'.

Trusses can be two-dimensional (planar) or three-dimensional (prismatic).

Prismatic or space-trusses linked together become space frames.

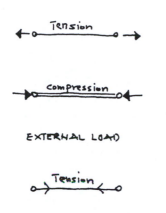

Tension

Compression

EXTERNAL LOAD

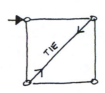

Tension

Compression

INTERNAL FORCE

TRUSS SINGLE ELEMENTS

STABLE FOR
ONE LOAD
DIRECTION

STABLE FOR
BOTH LOAD
DIRECTIONS

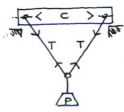

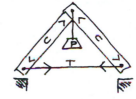

FORCES IN THE SIMPLEST TRUSS FORM

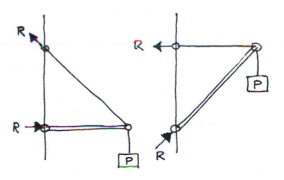

TRIANGULATION AND FORCES

UNSTABLE          STABLE

Minimum number of members for stability  = 3 linked together

55

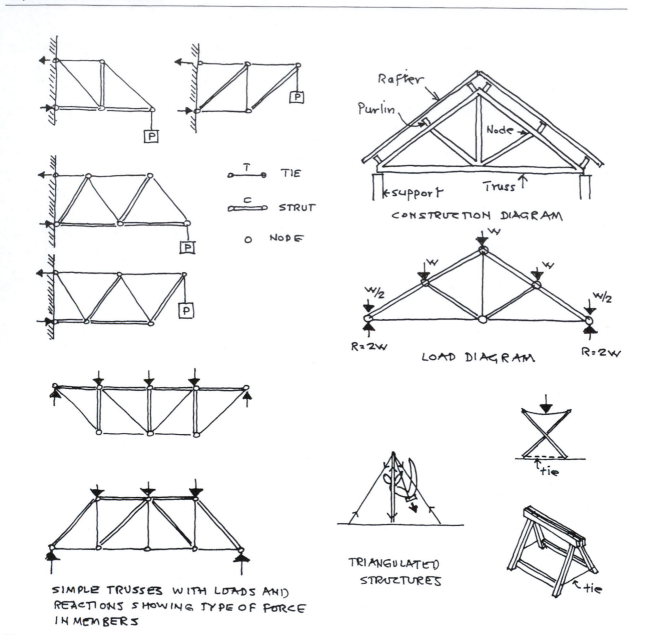

T    TIE

C    STRUT

O    NODE

CONSTRUCTION DIAGRAM

Rafter

Purlin

Node

support    Truss

LOAD DIAGRAM

R=2W    R=2W

w    w    w/2    w/2

SIMPLE TRUSSES WITH LOADS AND
REACTIONS SHOWING TYPE OF FORCE
IN MEMBERS

TRIANGULATED
STRUCTURES

tie

tie

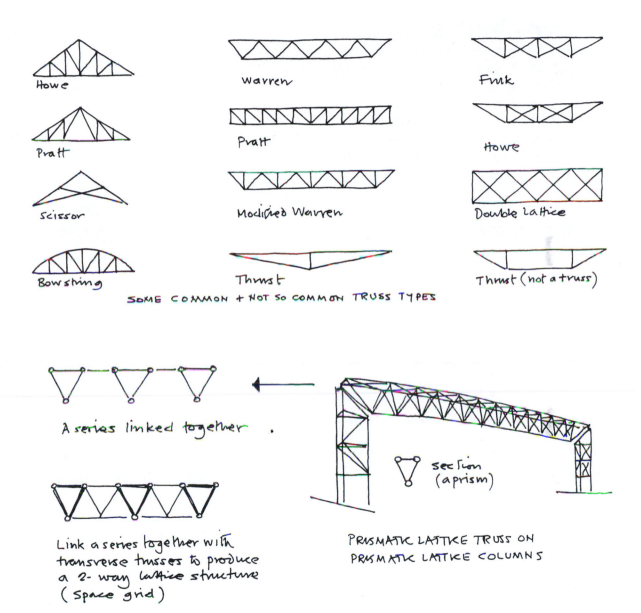

Howe

Pratt

Scissor

Bowstring

Warren

Pratt

Modified Warren

Thrust

Fink

Howe

Double Lattice

Thrust (not a truss)

SOME COMMON + NOT SO COMMON TRUSS TYPES

A series linked together

Link a series together with transverse trusses to produce a 2- way lattice structure (space grid)

section (a prism)

PRISMATIC LATTICE TRUSS ON PRISMATIC LATTICE COLUMNS

57

# Single-layer lattice grids

Sometimes called single-layer space frames, they are latticed structures with their structural action enhanced by folding. They span longitudinally instead of transversely and are capable of covering quite large areas.

Each fold line acts as a support edge interacting with adjacent planes to prevent deformation.

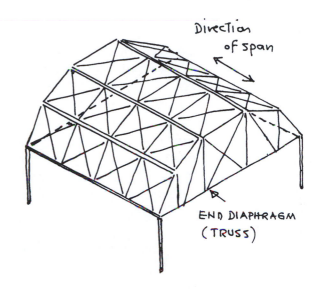

Direction of span

END DIAPHRAGM
(TRUSS)

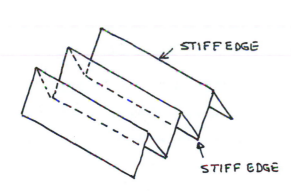

STIFF EDGE

STIFF EDGE

THE PRINCIPLE OF FOLDING
TO PROVIDE STIFFNESS

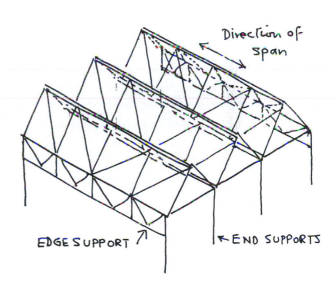

Direction of span

EDGE SUPPORT

END SUPPORTS

# Frameworks

Frameworks are composed of elements which when assembled in two or three directions form a skeletal structure.

The stiffness of a frame depends on the stiffness of the elements and the type of joints between frame members which can be pinned, fixed or partially fixed.

Pinned joint frames are unstable under load and require the addition of a further element to give stiffness: diagonal bracing or stiff panels.

Partially or fully fixed joint frames are stable under load. Loads on beam members cause deflection of the member and rotation of the adjacent joints. This rotation in turn causes deformation of the connected column members and in multi-member frames it becomes a complex problem to analyse. It is now usually solved by computer as the frame is statically indeterminate, i.e. the problem cannot be solved by simple calculations due to the interaction of one member with another through continuity at joint connections.

PLANE FRAMES

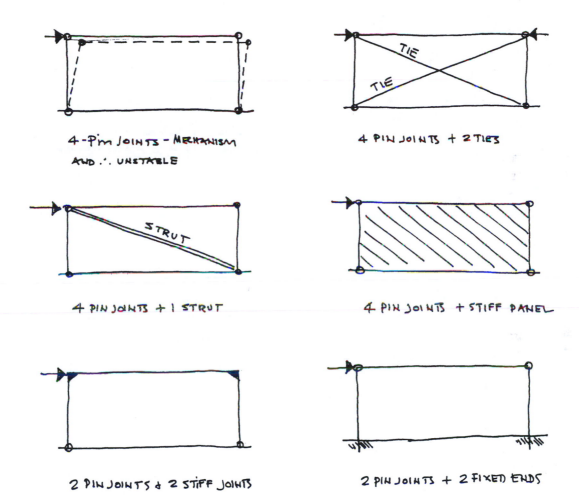

4-PIN JOINTS - MECHANISM AND ∴ UNSTABLE

4 PIN JOINTS + 2 TIES

4 PIN JOINTS + 1 STRUT

4 PIN JOINTS + STIFF PANEL

2 PIN JOINTS + 2 STIFF JOINTS

2 PIN JOINTS + 2 FIXED ENDS

## BASIC BRACING SYSTEMS

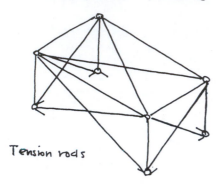

Tension rods

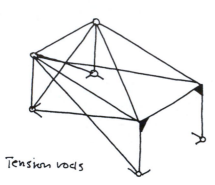

Tension rods

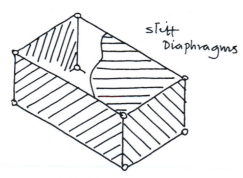

slift
Diaphragms

MINIMUM – 3 FACES + TOP BRACED

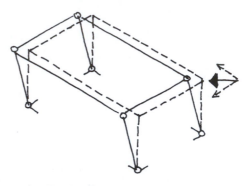

COLLAPSE DIAGRAM OF ALL PIN–JOINTED
FRAME WHICH IS A MECHANISM

BRACING RULES

Assume base is a rigid plane

Brace top plane

Brace any 3 vertical planes

Use diagonals, stiff diaphragms

or a combination of both

SOME EXAMPLES OF BRACING ALTERNATIVES FOR MULTI-BAYS

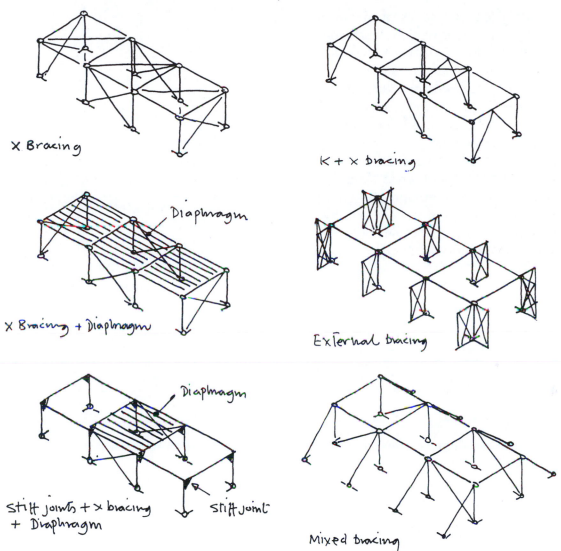

X Bracing

K + x bracing

Diaphragm

X Bracing + Diaphragm

External bracing

Diaphragm

Stiff joints + x bracing + Diaphragm

Stiff joint

Mixed bracing

# Grids

Grids are composed of a series of members arranged at right angles to one another, either parallel to the boundary supports (rectangular grids) or at 45° to the boundary supports (skew grids or diagrids). They behave structurally by load-sharing according to the position and direction of the members close to and further from the position of the load.

The structural analysis of such grids is complicated due to the number of variables involved and therefore is ideal for solution by computer. Grids are commonly used only for large spans where scale economies balance cost and construction complications.

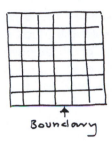

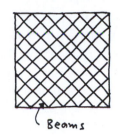

Boundary

Beams

RECTANGULAR GRID
( orthogonal )

SKEW GRID
( Diagrid )

STEEL OR ALLOY

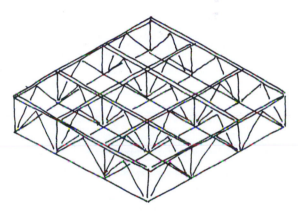

LATTICE GRID

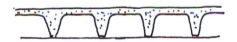

CONCRETE 2-WAY WAFFLE

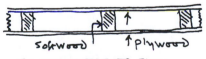

softwood     ↑plywood

TIMBER STRESSED SKIN

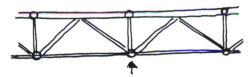

2-WAY LATTICE TRUSSES

65

# Space frames

Space frames are three-dimensional lattice structures made up from linked pyramids or tetrahedra into a two-layer or three-layer triangulated framework. Load span and edge conditions determine the form and depth of the space frame. Because of the continuous member linking, optimum load-sharing occurs and for large clear spans – above about 20m, the space frame is a very efficient form of structure with a span/depth ratio of approximately 20:1.

Plan proportions should be near square and not exceed 1.5:1.

THE BASIC UNIT

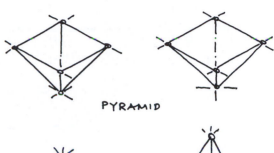

PYRAMID

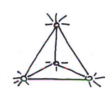

TETRA HEDRON

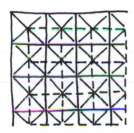
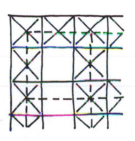
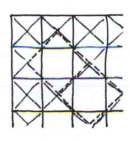
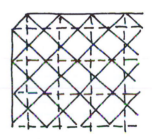

SQUARE ON SQUARE OFFSET

SQUARE ON LARGER SQUARE

SQUARE ON DIAGONAL

DIAGONAL ON SQUARE

TYPICAL GRID ARRANGEMENTS

SOME SPACE FRAME SYSTEMS !

NODUS   SPACE DECK   MERO   TRIODETIC   OKTAPLATTE   UNISTRUT   SDC

↑ OUT OF PRODUCTION

67

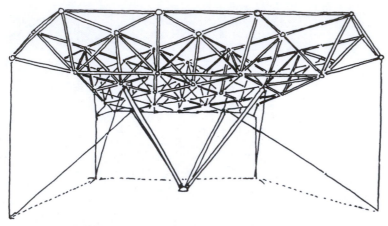

SPACE FRAME - LONDON ZOO
Centrally supported with tension cables for stability

CABLE + MAST SUPPORTED SPACE FRAME
AS ENTRANCE CANOPY

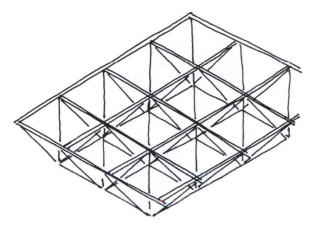

SQUARE ON SQUARE

SQUARE ON LARGER SQUARE

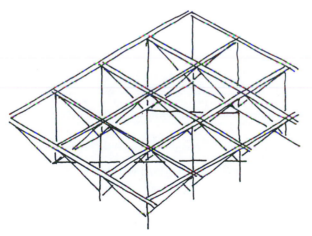

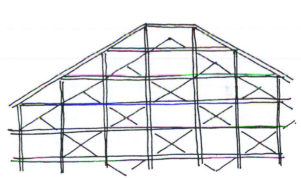

SQUARE ON DIAGONAL

DIAGONAL ON SQUARE

SPACE FRAME GRIDS

# Plates

Plates or flat slabs are generally horizontal elements with a length and breadth which are large in comparison with their thickness – span to depth up to 40:1. They are designed to span in two directions at right angles and may be flat, have stiffening strips or thickening at supporting column points.

Multi-bay plates are statically indeterminate and are calculated by textbook design factors or computer.

Slabs can be designed around lines of equal stress but formwork is elaborate and thus expensive (cf the work of Pier Luigi Nervi).

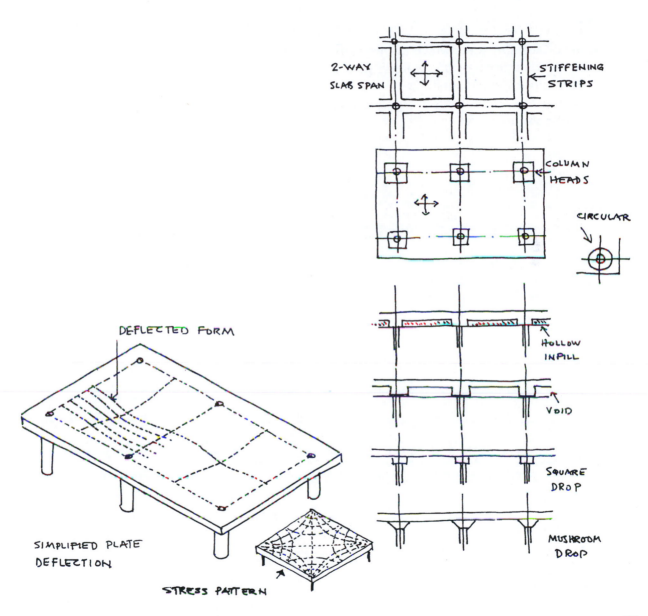

2-WAY
SLAB SPAN

STIFFENING
STRIPS

COLUMN
HEADS

CIRCULAR

HOLLOW
INFILL

VOID

SQUARE
DROP

MUSHROOM
DROP

DEFLECTED FORM

SIMPLIFIED PLATE
DEFLECTION

STRESS PATTERN

# Shells

Shells are surface structures which are curved in one of two directions or are warped as in the hyperbolic paraboloid shell.

Structural forces in shells are largely pure tension and compression.

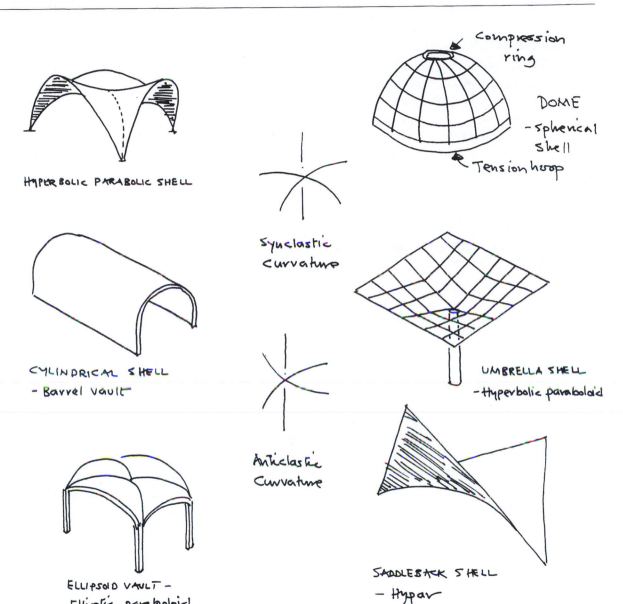

HYPERBOLIC PARABOLIC SHELL

Compression ring

DOME
- spherical shell

Tension hoop

Synclastic Curvature

CYLINDRICAL SHELL
- Barrel vault

UMBRELLA SHELL
- Hyperbolic paraboloid

Anticlastic Curvature

ELLIPSOID VAULT -
Elliptic paraboloid

SADDLEBACK SHELL
- Hypar

73

# Stressed skins

A combination of thin plates with rib-stiffeners is a stressed skin surface.

The ribs contribute stiffness to what would otherwise be a too thin and flexible sheet material, which under load would buckle.

The material used for stressed skin construction can be metal, timber, GRP, or sometimes a combination (e.g. 'Nomex' – see composite material panels).

Analysis of stressed skins is carried out by computer or by testing, according to the complexity of the problem since this structure is statically indeterminate.

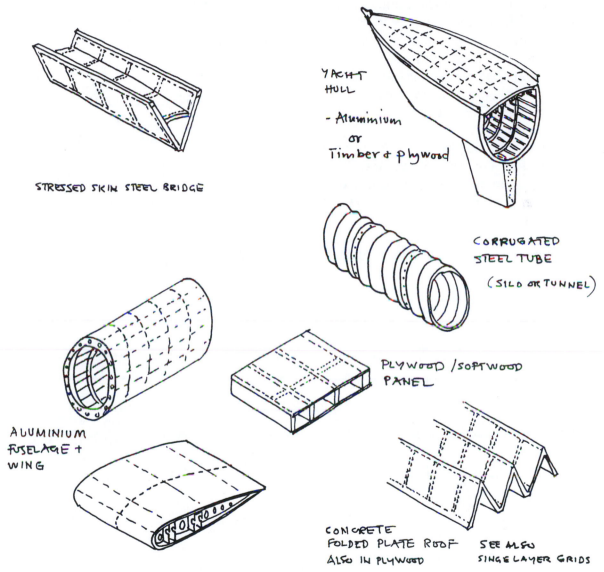

STRESSED SKIN STEEL BRIDGE

YACHT
HULL

- Aluminium
  or
Timber & plywood

CORRUGATED
STEEL TUBE

( SILO OR TUNNEL )

PLYWOOD / SOFTWOOD
PANEL

ALUMINIUM
FUSELAGE +
WING

CONCRETE
FOLDED PLATE ROOF
ALSO IN PLYWOOD

SEE ALSO
SINGLE LAYER GRIDS

75

# Membranes

In membrane structures all the primary forces are arranged to be in tension, either in the form of cables forming a net or by means of a coated fabric with tensioned edge cables. Loads from the membrane can be taken to the ground via compression masts with perimeter anchor cables or by some other form of aerial structure.

Stress concentrations tend to occur at the boundaries and curved cables are often introduced to even them out (tear drops and zigzags at the mast top, boundary cables at the edge). Curvature of the surface must be maintained in two directions (anticlastic) otherwise flutter will occur under wind load and failure may result.

Pneumatics are air supported membranes usually without any other form of structure required to support them, except a foundation ring beam to act as an anchor.

There are a number of fabric types in use and others constantly under development. The three typical ones in current use are:

Polyvinyl chloride coated polyester – PVC polyester
Polytetrafluoroethylene coated glass fibre – PTFE glass
Ethylene-tetra-fluoroethylene foil – ETFE foil

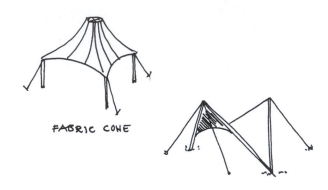

FABRIC CONE

HYPAR FABRIC TENT

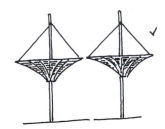

INVERTED FABRIC CONES
STEEL MASTS

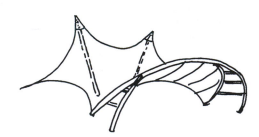

MAST + ARCH SUPPORTED MEMBRANE

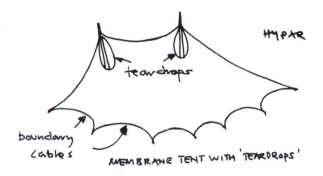

HYPAR

teardrops

boundary cables

MEMBRANE TENT WITH 'TEARDROPS'

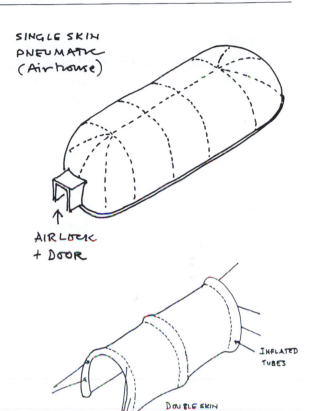

SINGLE SKIN PNEUMATIC (Air house)

AIR LOCK + DOOR

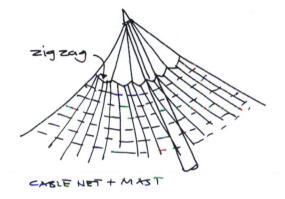

zig zag

CABLE NET + MAST

INFLATED TUBES

DOUBLE SKIN PNEUMATIC (self supporting)

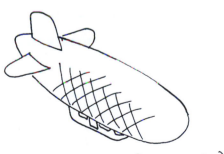

DIRIGIBLE (Gas filled airship)

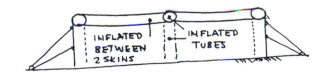

INFLATED BETWEEN 2 SKINS

INFLATED TUBES

77

# Hybrids

There are a number of structural types which do not fit into any of the four previous classifications and these are defined as hybrids. It is a fact that although the primary type may be 'solid', 'skeletal' etc., secondary elements of a different type may be part of the structure. The hybrid is where there is a combination of two types of near equal dominance.

Many tension-assisted structures fall into this category and typically will consist of the following combinations:

*Steel and tensile membrane*
*Structural glass and steel*
*Masonry and steel*
*Timber/plastic and steel*

Examples of the above, in order, are as follows:

Schlumberger Cambridge Research – Cambridge
Waterloo International Station Concourse Wall
Pabellón de Futur–Seville Expo
IBM Travelling Exhibition

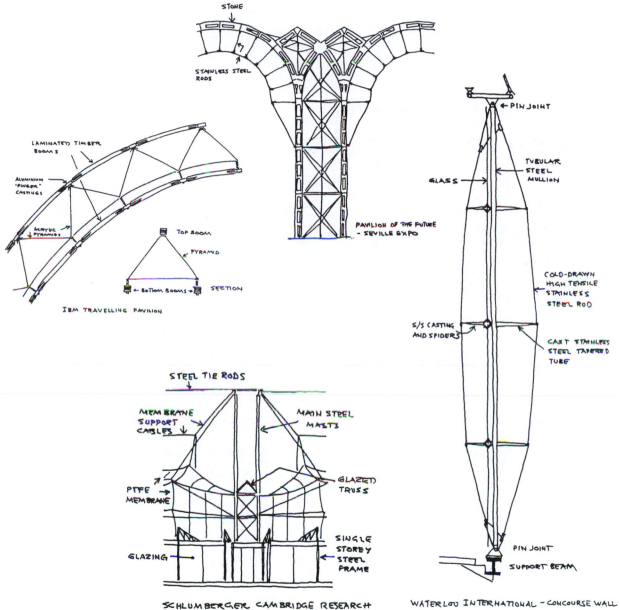

STONE

STAINLESS STEEL RODS

PAVILION OF THE FUTURE – SEVILLE EXPO

LAMINATED TIMBER BOOMS

ALUMINIUM 'FINGER' CASTINGS

ACRYLIC PYRAMIDS

TOP BOOM

PYRAMID

← BOTTOM BOOMS →    SECTION

IBM TRAVELLING PAVILION

PIN JOINT

TUBULAR STEEL MULLION

GLASS →

COLD-DRAWN HIGH TENSILE STAINLESS STEEL ROD

S/S CASTING AND SPIDERS

CAST STAINLESS STEEL TAPERED TUBE

STEEL TIE RODS

MEMBRANE SUPPORT CABLES ↓

MAIN STEEL MASTS

GLAZED TRUSS

PTFE MEMBRANE

GLAZING →

SINGLE STOREY STEEL FRAME

PIN JOINT

SUPPORT BEAM

SCHLUMBERGER CAMBRIDGE RESEARCH

WATERLOO INTERNATIONAL – CONCOURSE WALL

79

# 8

# Some further significant structures and assemblies

A detailed look at the work of well-known engineers throws up a number of original and inventive solutions for both complete structural assemblies and structural parts.

As with hybrids in Chapter 7, these designs are difficult to classify. They are always a combination of, at least, tension and compression. They can, however, be put into categories and these are listed with the relevant structures illustrated.

*Primary tension*

Kempinski Hotel, Munich

ETFE Foil 'Pillow' – Eden Project

Parc de la Villette, Paris –
Cable Trussed Glass Wall

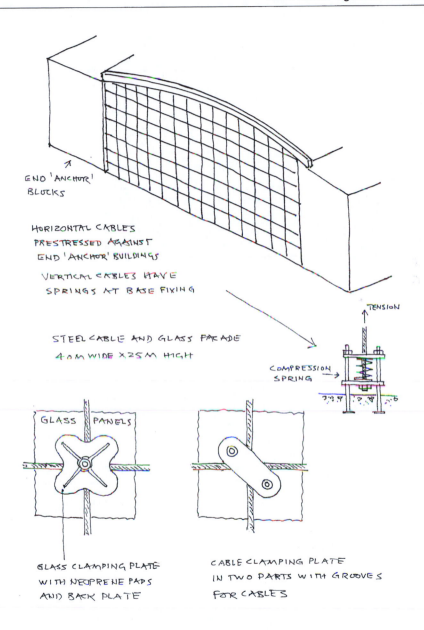

END 'ANCHOR' BLOCKS

HORIZONTAL CABLES
PRESTRESSED AGAINST
END 'ANCHOR' BUILDINGS

VERTICAL CABLES HAVE
SPRINGS AT BASE FIXING

STEEL CABLE AND GLASS FACADE
40M WIDE X 25M HIGH

TENSION

COMPRESSION
SPRING

GLASS PANELS

GLASS CLAMPING PLATE
WITH NEOPRENE PADS
AND BACK PLATE

CABLE CLAMPING PLATE
IN TWO PARTS WITH GROOVES
FOR CABLES

KEMPINSKI HOTEL - MUNICH - CABLE / GLASS WALL

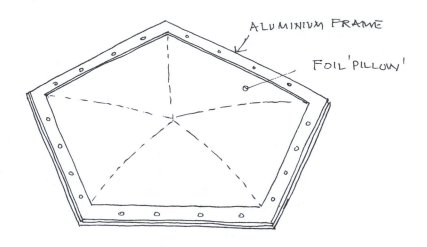

ALUMINIUM FRAME

FOIL 'PILLOW'

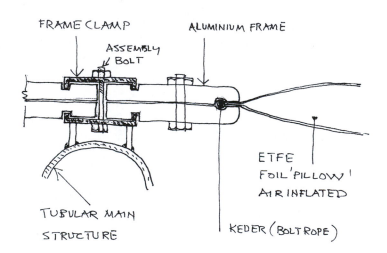

FRAME CLAMP

ASSEMBLY BOLT

ALUMINIUM FRAME

TUBULAR MAIN STRUCTURE

ETFE FOIL 'PILLOW' AIR INFLATED

KEDER (BOLT ROPE)

'EDEN PROJECT - CORNWALL
ETFE FOIL PILLOW AND ALUMINIUM CLADDING

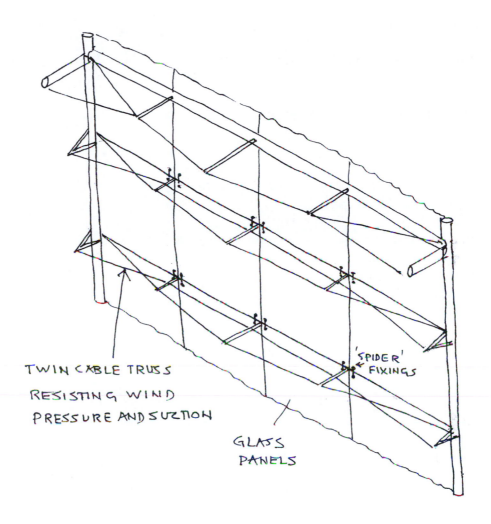

TWIN CABLE TRUSS
RESISTING WIND
PRESSURE AND SUCTION

'SPIDER' FIXINGS

GLASS PANELS

PARC de la VILLETTE - PARIS

CABLE TRUSSED GLASS WALL

## Tension and compression

The Skylon – Festival of Britain 1951

Visionary Structures – Robert le Ricolais

Hong Kong Aviary

International Conference Centre, Paris – fully adjustable main glazing bracket

Stuttgart 21 – Station Roof Structure

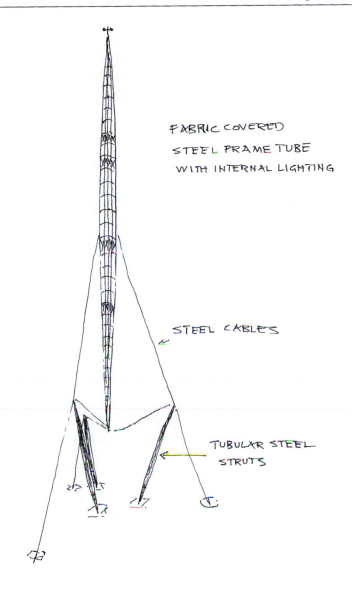

FABRIC COVERED
STEEL FRAME TUBE
WITH INTERNAL LIGHTING

STEEL CABLES

TUBULAR STEEL
STRUTS

THE SKYLON – FESTIVAL OF BRITAIN · LONDON 1951

AN EARLY EXAMPLE OF A TENSEGRITY STRUCTURE

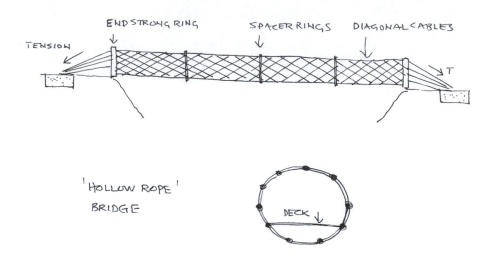

TENSION    END STRONG RING    SPACER RINGS    DIAGONAL CABLES    T

'HOLLOW ROPE'
BRIDGE

DECK ↓

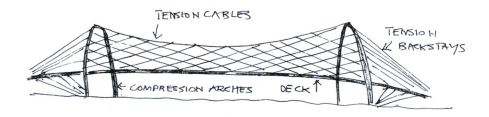

TENSION CABLES

TENSION BACKSTAYS

COMPRESSION ARCHES    DECK ↑

'HOLLOW ROPE' SUSPENSION BRIDGE

TWO EXAMPLES OF ROBERT LE RICOLAIS'
'VISIONARY' STRUCTURES

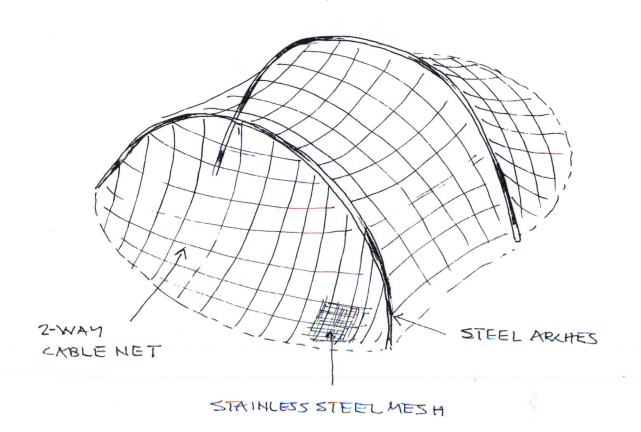

2-WAY
CABLE NET

STAINLESS STEEL MESH

STEEL ARCHES

HONG KONG AVIARY

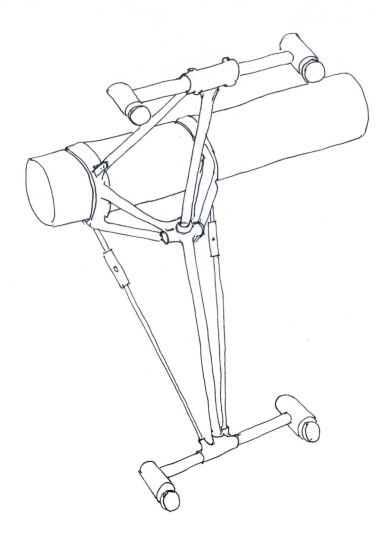

INTERNATIONAL CONFERENCE CENTRE – PARIS

FULLY ADJUSTABLE MAIN GLAZING BRACKET

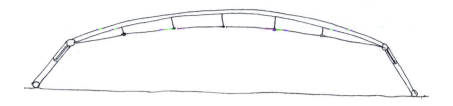

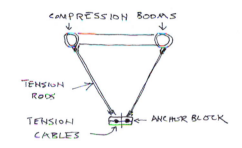

COMPRESSION BOOMS

TENSION RODS

TENSION CABLES

ANCHOR BLOCK

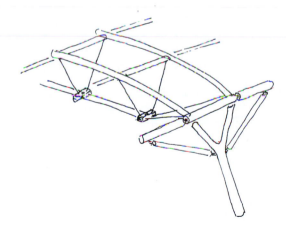

STUTTGART 21 — STATION ROOF STRUCTURE

*Primary compression*

Thin Concrete Shells – Switzerland

Geodesic Domes – Eden Project,
Cornwall

Compression Arch and Joint –
National Botanic Garden of Wales

STIFFENING
'EYEBROWS'

CONCRETE SHELL - GARDEN CENTRE · PARIS

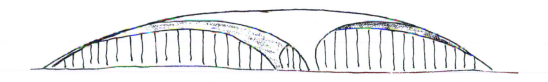

KILCHER FACTORY near SOLOTHURN - CONCRETE SHELL

EXAMPLES OF EXTREMELY THIN CONCRETE SHELLS

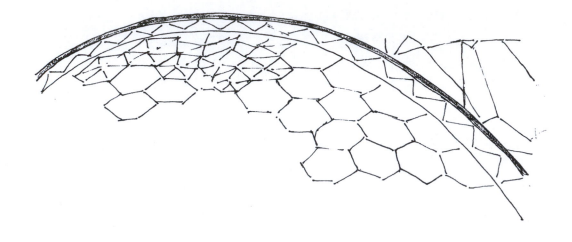

EDEN PROJECT - CORNWALL

PART OF A MAIN GEODESIC DOME
AND SUPPORTING ARCH

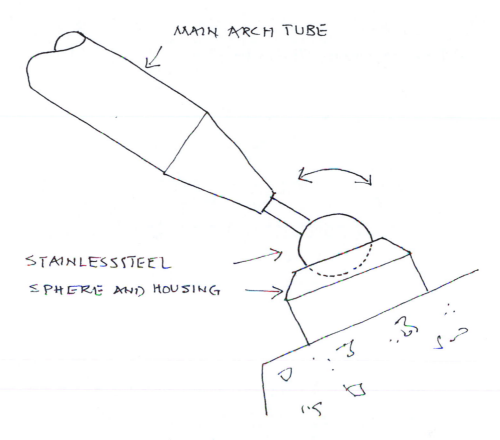

MAIN ARCH TUBE

STAINLESS STEEL
SPHERE AND HOUSING

NATIONAL BOTANIC GARDEN OF WALES

MAIN ARCH BEARING IN STAINLESS STEEL

IT ALLOWS ROTATION DUE TO TEMPERATURE

VARIATION [THERE IS NO UPLIFT!]

## *Composite*

### DuPont Competition – 'Zip-Up' House

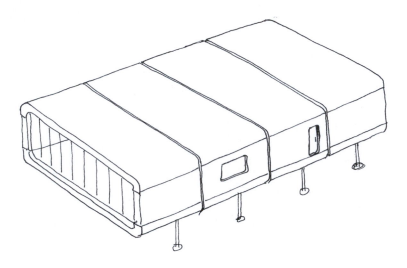

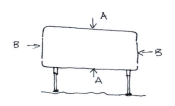

TYPE A – ROOF and FLOOR
TYPE B – WALLS

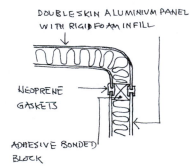

DOUBLE SKIN ALUMINIUM PANEL
WITH RIGID FOAM INFILL

NEOPRENE
GASKETS

ADHESIVE BONDED
BLOCK

DUPONT COMPETITION – 'ZIP-UP' HOUSE [1966]

AN EARLY EXAMPLE OF MONOCOQUE REPETITION
FOR BATCH PRODUCTION

# Appendices

## Appendix I: Tensile strength of some common materials

| Material | Tensile strength – S | | |
|---|---|---|---|
| | lb/in² | MN/m² | Relative strength |
| Cement and concrete | 600 | 4.1 | 0.6 |
| Ordinary brick | 800 | 5.5 | 0.8 |
| Fresh tendon (animal) | 12,000 | 82 | 12 |
| Hemp rope | 12,000 | 82 | 12 |
| Wood (air dry) along grain | 15,000 | 103 | 15 |
| Wood (air dry) across grain | 500 | 3.5 | 0.5 |
| Fresh bone | 16,000 | 110 | 16 |
| Ordinary glass | 5,000–25,000 | 35–175 | 5–25 |
| Human hair | 28,000 | 192 | 28 |
| Spider's web | 35,000 | 240 | 35 |
| Good ceramics | 5,000–50,000 | 35–350 | 5–50 |
| Silk | 50,000 | 350 | 50 |
| Cotton fibre | 50,000 | 350 | 50 |
| Catgut | 50,000 | 350 | 50 |
| Flax | 100,000 | 700 | 100 |
| Glassfibre plastics | 50,000–150,000 | 350–1,050 | 50–150 |
| Carbon fibre plastics | 50,000–150,000 | 350–1,050 | 50–150 |
| Nylon thread | 150,000 | 1,050 | 150 |

## Tensile strength of some common materials (cont'd)

| Material | Tensile strength–S | | |
|---|---|---|---|
| | lb/in² | MN/m² | Relative strength |
| Steel piano wire (very brittle) | 450,000 | 3,100 | 450 |
| High tensile engineering steel | 225,000 | 1,550 | 225 |
| Commercial mild steel | 60,000 | 400 | 60 |
| Traditional wrought iron | 15,000–40,000 | 100–300 | 15–40 |
| Traditional cast iron (very brittle) | 10,000–20,000 | 70–140 | 10–20 |
| Modern cast iron | 20,000–40,000 | 140–300 | 20–40 |
| Aluminium: cast | 10,000 | 70 | 10 |
| Aluminium: wrought alloys | 20,000–80,000 | 140–600 | 20–80 |
| Magnesium alloys | 30,000–40,000 | 200–300 | 30–40 |
| Titanium alloys | 100,000–200,000 | 700–1,400 | 100–200 |
| Carbon fibre (high strength) | 270,000 | 4,000 | 270 |
| Kevlar 49 | 270,000 | 4,000 | 270 |

# *E* values for some common materials

| Material | E value | | |
|---|---|---|---|
| | lb/in² | MN/m² | Relative stiffness |
| Rubber | 1,000 | 7 | 1 |
| Shell membrane of an egg | 1,100 | 8 | 1.1 |
| Human cartilage | 3,500 | 24 | 3.5 |
| Human tendon | 80,000 | 600 | 80 |
| Wallboard | 200,000 | 1,400 | 200 |
| Unreinforced plastics, polythene and nylon | 200,000 | 1,400 | 200 |
| Plywood | 1,000,000 | 7,000 | 1,000 |
| Wood (along grain) | 2,000,000 | 14,000 | 2,000 |
| Fresh bone | 3,000,000 | 21,000 | 3,000 |
| Magnesium metal | 6,000,000 | 42,000 | 6,000 |
| Ordinary glass | 10,000,000 | 70,000 | 10,000 |
| Aluminium alloys | 10,000,000 | 70,000 | 10,000 |
| Brasses and bronzes | 17,000,000 | 120,000 | 17,000 |
| Kevlar 49 | 19,000,000 | 130,000 | 19,000 |
| Iron and steel | 30,000,000 | 210,000 | 30,000 |
| Carbon fibre (high strength) | 60,000,000 | 420,000 | 60,000 |
| Aluminium oxide (sapphire) | 60,000,000 | 420,000 | 60,000 |
| Diamond | 170,000,000 | 1,200,000 | 170,000 |

# Appendix II: Bending and deflection formulae for beams

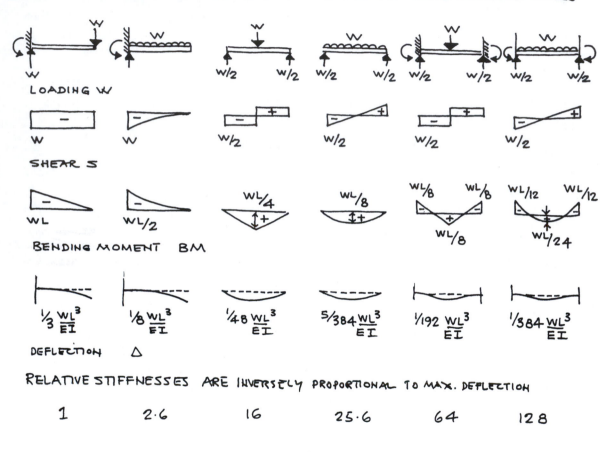

SHEAR BENDING AND DEFLECTION DIAGRAMS FOR SOME STANDARD CASES

LOADING W

SHEAR S

BENDING MOMENT BM

DEFLECTION △

RELATIVE STIFFNESSES ARE INVERSELY PROPORTIONAL TO MAX. DEFLECTION

| 1 | 2·6 | 16 | 25·6 | 64 | 128 |
|---|---|---|---|---|---|

RELATIVE STRENGTHS ARE INVERSELY PROPORTIONAL TO MAX. BENDING MTS.

| 1 | 2 | 4 | 8 | 8 | 12 |
|---|---|---|---|---|---|

# Appendix III: Reading list

Bill Addis, *The Art of the Structural Engineer*,
Artemis
Some philosophy and a wide range of recent case studies of building structure in all materials

Allen Andrews, *Back to the Drawing Board: The Evolution of Flying Machines*,
David & Charles
Drawings, models, photographs and text from before the birth of the lighter-than-air machine to nearly the present. A must for anyone interested in aircraft development

Fred Angerer, *Surface Structures in Building*,
Alec Tiranti
Probably the best non-mathematical book on surface structures, with excellent diagrams

Cecil Balmond with Jannuzzi Smith, *'Informal'*,
Prestel
A book by an engineer who collaborates with architects in creating 'non-cartesian' free-form structures. The book is part engineering, part mathematics and part philosophy and outlines a different approach to structural thinking

Derrick Beckett, *Bridges*, Paul Hamlyn
A very good general view

Behnisch/Hartung, *Elsenconstruktionen Des 19 Jahrhunderts* (in German), Schiriner/Mosel
Comprehensive coverage of nineteenth-century iron and steel engineering and architecture

Adriaan Beukers and Ed van Hinte, *Lightness – the Inevitable Renaissance of Minimum Energy Structures*,
010 publishers, Rotterdam
Essential reading for any designer interested in ways of using 'smart' materials in the most economical way. It has some very revealing examples and statistics. Based on research carried out at the Faculty of Aerospace Engineering, Delft Technical University

John Borrego, *Space Grid Structures*,
MIT Press
A comprehensive catalogue of three-dimensional structures, with good diagrams and photographs of models

Alan J Brookes and Chris Grech, *The Building Envelope*, Butterworth Architecture

Isambard Kingdom Brunel, *Recent Works*,
Design Museum
Analysis and drawings by practising architects and
engineers of a number of Brunel's famous works
including the Royal Albert Bridge at Saltash, the
*Great Eastern* and Paddington Station

Santiago Calatrava, *The Daring Flight*, Electa

Centre George Pompidou/Le Moniteur, *l'art de
l'ingénieur*
A biography of engineering and engineers worldwide. Only published in French

Connections – *Studies in Building Assembly*,
Butterworth Architecture
Two excellent books on the 'parts of buildings' with
clear drawings to complement the photographs

Keith Critchlow, *Order in Space*, Thames and
Hudson
The book on three-dimensional geometry

Christopher Dean (Ed.), *Housing the Airship*,
Architectural Association

James Dyson, *Against the Odds*
Orion Business Books
James Dyson's fascinating autobiography outlining
his design and business tribulations from Royal
College of Art days to the final production of the
Dyson Cyclone vacuum cleaner

H Engel, *Structure Systems*, Penguin

Giunti Florence, *The Art of Invention*
Leonardo and Renaissance Engineers

J E Gordon, *Structures and Why Things Don't Fall
Down*, Penguin
*The New Science of Strong Materials*, Penguin
*The Science and Structure of Things*, Scientific
American Library
Three books that are essential reading for an under-
standing of general structures, with very little maths

Sembach Leuthäuser Gössel, *Twentieth Century
Furniture Design*, Taschen
Probably the most comprehensive book on the
subject

Erwin Heinle and Fritz Leonhardt, *Türme (Towers)*
(in German), DVA
Wide coverage of towers worldwide through the
ages

Monica Henning-Schefold, *Transparanz und
Masse*, Du Mont
A German book illustrating glazed malls and halls
from 1800 to 1880 – masses of photographs

John Hix, *The Glass House*, Phaidon
A review of glass architecture up to the present

Alan Holgate, *The Work of Jorg Schlaich and his
Team*,
Edition Axel Menges
A comprehensive and beautifully produced book on
the work of the German engineer who to my mind is
one of the great inspirational engineers of the twentieth century and who, together with his team, is still
working today

Institution of Structural Engineers, *Structural Use of Glass in Buildings*
Probably the best technical design guide to structural glass and glazing including many design examples

Ross King, *Brunelleschi's Dome,*
Chatto & Windus
The story of the construction of the Great Cathedral in Florence (completed in 1436)

Fritz Leonhardt, *Brüken (Bridges)*, Architectural Press
The comprehensive coverage of bridges worldwide through the ages

Angus J MacDonald, *Structure and Architecture,* Butterworth Architecture
This could be considered to be the companion volume to *Tony Hunt's Structures Notebook*

Rowland Mainstone, *Developments in Structural Form*, 2nd Edition, Allen Lane
Probably the best and most comprehensive book on structures of all ages with marvellous photographic coverage

Z Makowski, *Steel Space Structures*, Michael Joseph
A very good review of built structures with excellent photographs and diagrams

Robert W Marks, *The Dymaxion World of Buckminster Fuller*, Reinhold
The best of Bucky Fuller's ideas

Meadmore, *The Modern Chair*, Studio Vista
A good, but not very comprehensive review of modern chairs with scale drawings

John and Marilyn Newhart and Ray Eames, *Eames Design*, Ernst & John
A complete record of the multi-faceted work of Charles and Ray Eames

Frei Otto, *Tension Structures*, Volumes I and 2, MIT Press
These and other later books cover nets, membranes and pneumatics and see also the IL series

Martin Pearce and Richard Jobson, *Bridge Builders*, Wiley-Academy
Recent book on bridges again with superb photographs and illustrations, some overlap with Matthew Wells' book and with a briefer text

Jean Prouvé, Prefabrication: *Structures and Elements*, Pall Mall Press
Prouvé was a much underrated designer whose inventive work in the field of lightweight panels and structures has never been bettered although much of it was carried out fifty years ago

Peter Rice, *An Engineer Imagines*,
Artemis
The engineering and philosophical memoirs of the famous engineer who sadly died much too young. A number of the seminal structures of the twentieth century are here including the Sydney Opera House, the Centre Pompidou, the Lloyds Building and many others.

Lyall Sutherland, *Master of Structure: Engineering Today's Innovative Buildings*
Publ. Lawrence King.
This is the book that gives credit to the usually unacknowledged work of the engineer. It clearly shows the collaboration between engineer and architect that is essential to produce a good building

Robert le Ricolais, *Visions and Paradox*,
Fundacion Cultural C.O.A.M.
The inventor of the 'Hollow Rope' structural principle with drawings and illustrations of all his many experimental models

Salvadori and Heller, *Structure in Architecture*, Prentice Hall
Salvadori and Levi, *Structural Design in Architecture*, Prentice Hall
Two very good books on building structures, the first entirely non-mathematical, the second with worked examples

Daniel Schodek, *Structures*, Prentice Hall
A good comprehensive textbook on basic principles with analysis and design

Dennis Sharp (Ed.), *Santiago Calatrava*, Book Art

Thomas Telford Press, *The Engineers Contribution to Contemporary Architecture*
Monographs by various authors on the following engineers: Eladio Dieste, Anthony Hunt, Heinz Isler, Peter Rice and Owen Williams. A self-explanatory series of books reflecting the title, with more volumes to come

Eduardo Torroja, *Philosophy of Structures*, University of California Press

Maritz Vandenburg, *Soft Canopies*, Academy Editions
A good primer on tensile membrane structures

Maritz Vandenburg, *Glass Canopies and Cable Nets*, Academy Editions
Two good primers on the subjects with beautiful drawings.

Konrad Wachsmann, *Turning Point in Building*, Reinhold
A seminal book on jointing and ideas on long span structures

Matthew Wells, *30 Bridges*, Lawrence King
A book discussing the history of bridge building and giving examples together with superb sketches and photographs of 30 interesting bridges with analysis of their behaviour

Michael White, *Isaac Newton the Last Sorcerer*, 4th Estate
A fascinating biography of a brilliant but not entirely likeable genius

Chris Wilkinson, *Supersheds*, 2nd Edition, Butterworth Architecture
The definitive work on clear span structures from the nineteenth century to the present